CW01669914

Science

CHEMISTRY

FOR COMMON ENTRANCE

Answers

Ron Pickering

GALORE PARK
AN HACHETTE UK COMPANY

About the author

Ron Pickering has published a number of very successful books covering the GCSE, IGCSE and A level syllabi and has worked in both maintained and independent education for more than 30 years. He now divides his time between teacher training, both in the UK and overseas, and writing, and has been a science advisor and curriculum manager at Altrincham Grammar School for Girls, as well as a Science Inspector for OFSTED.

Ron extends his interest in science by spending many hours photographing animals, both in the wild and in captive environments, and tries to maintain some level of fitness by off-road cycling.

Orders: please contact Bookpoint Ltd, 130 Milton Park, Abingdon, Oxon OX14 4SB. Telephone: +44 (0)1235 827827. Lines are open 9.00a.m. 5.00p.m., Monday to Saturday, with a 24-hour message answering service. Visit our website at www.galorepark.co.uk for details of other revision guides for Common Entrance, examination papers and Galore Park publications.

Published by Galore Park Publishing Ltd

An Hachette UK company

Carmelite House, 50 Victoria Embankment, London, EC4Y 0DZ

www.galorepark.co.uk

Impression number 10 9 8 7 6 5 4 3 2 1

2019 2018 2017 2016 2015

Typeset in 11.5/13 ITC Officina Sans/Book by Integra Software Services Pvt. Ltd, Pondicherry, India.

Printed in the UK.

New illustrations by Integra Software Services Pvt. Ltd, Pondicherry, India.

Some illustrations by Graham Edwards were re-used. The publishers will be pleased to make the necessary arrangements with regard to these illustrations at the first opportunity.

A catalogue record for this title is available from the British Library.

ISBN: 9781471847134

Contents

Introduction

About this book

Science for Common Entrance: Chemistry covers the Chemistry component of Science at Key Stage 3 and is part of an ISEB-approved course leading to 13+ Common Entrance. In this book you will find all the answers to the Exercise questions and Investigation questions contained in the Chemistry book. These are presented by chapter, as in the Student's Book. Answers to all of the Exercises are given first, followed by the answers to Investigation questions.

The answers given here should be seen as a guide only. Students of different abilities will approach and respond to questions differently, and credit should be given as appropriate.

Investigations in science

Exercise 1: Made to measure

1 Glass and plastic are useful materials because they:

- are transparent
- can be coloured
- can be cut to shape
- are rigid.

2 Glass is more useful than plastic when making measuring equipment because the scale cannot be damaged by scratch marks and liquids can be heated in glass beakers.

3 Plastic is more useful than glass when making measuring equipment because it is easier to mould into shape and is less likely to break.

Extension questions

4 From the measurements you can tell that a certain volume of alcohol has less mass than the same volume of water.

Students may have an understanding of density from their physics studies: Alcohol has a lower density than water. Water has a density of 1.0 g/cm^2 and alcohol has a density of 0.8 g/cm^3.

5 (a) 140 g

(b) 140 ml – not quite enough if she normally drinks 200 ml in four hours.

Exercise 2: Experiments in chemistry

1 (a) Conical flask: used for mixing solutions, without heating

(b) Tripod: used to support apparatus above a Bunsen burner

(c) Measuring cylinder: used for measuring the volume of liquids

(d) Filter funnel: used to separate solids from liquids using filter paper

(e) Spatula: used for handling solid chemicals; for example when adding a solid to a liquid

(f) Pipette: used to measure and transfer small volumes of liquid

2 (a) 1220 g

(b) 259 s

(c) 22 min 0 s

(d) 2.34 litres

(e) 3.40 kg

(f) 2984 ml

Extension question

3 Here are some possible answers:

- Never carry out experiments in the laboratory without a teacher present.
- An open flame in a lab is a hazard. Never leave a lit Bunsen burner unattended.
- Looking into a vessel (or test tube) when heating something is dangerous; you should look in from the side, or heat substances in glass vessels so that the contents are clearly visible.
- Long hair should be tied back.
- Ties should be tucked in or removed.
- Laboratory coats should be fastened closed.
- Bare feet are dangerous, causing slipping and possible breakages of glass. You should wear sensible shoes in the lab.
- Spillages should be dealt with safely. Use gloves, a well-soaked cleaning rag and a plastic washing-up bowl with plenty of water.
- Eye protection (goggles) must be used when handling acids and indeed most solutions.
- Proper space must be allowed between each workstation; this lab is too crowded for safety.
- Large glass bottles should be stored on lower shelves and not with books.
- Fumes are dangerous; work in a fume cupboard when necessary.
- Do not leave objects (e.g. books and bags) lying on the floor; they can be dangerous obstructions.
- Dangers to the skin: wear gloves and overalls/labcoats when necessary.
- Do not carry test tubes or beakers of liquid around the laboratory; you may cause spillages.
- Do not run in the lab.
- Do not eat in the lab.
- Do not over reach for items.
- Do not carry too much at once.

Exercise 3: The Bunsen burner

1 (a) The independent variable in this experiment is whether the air hole is open or closed.

(b) The dependent variable in this experiment is the time taken for the water to boil.

(c) The following variables should be controlled:

(i) the size and shape of the beaker

(ii) the starting temperature of the water

(iv) the position of the Bunsen burner below the beaker

(v) the position of the gas tap (how much flow of gas)

(vi) the volume of water in the beaker.

2 (a) II and III cannot be compared in a fair test because there are three variables, the volume of water, the starting temperature and the gas flow.

(b) (i) Sara is testing whether an open or a closed air hole has a greater effect on the time it takes for the water to boil.

(ii) She **concludes** that the water takes less time to boil from her data; she can then **deduce/hypothesise** that the open air hole produces more thermal energy.

(c) (i) Sara is testing whether the starting temperature affects the time it takes the water to boil.

(ii) She can conclude that a higher starting temperature reduces the boiling time.

(d) Sara should compare the results I and V to find the effect of volume of water on the time taken for the water to boil.

Exercise 4: Testing

1 (a) Test using anhydrous copper sulfate or cobalt chloride paper.

(b) Anhydrous copper sulfate turns from blue to white if water is present. Cobalt chloride paper turns from blue to pink if water is present.

2 Limewater can be used to test for **carbon dioxide**. The limewater will turn **milky** if carbon dioxide is present. The gas called **oxygen** will make a glowing splint **relight**. A lighted splint will make the gas called **hydrogen** produce a sound like a **pop**.

1 Particle theory and states of matter

Exercise 1.1: Solids, liquids and gases

1 wood – solid; carbon dioxide – gas; snow – solid; plastic – solid; salt – solid; vinegar – liquid; stone – solid; lime juice – liquid; water vapour – gas; tomato ketchup – liquid

2

Does it ...?	Solid	Liquid	Gas
Melt	Yes		
Freeze		Yes	
Boil		Yes	
Compress		Not really	Yes
Conduct thermal energy	Yes (if a metal)		
Expand	Yes	Yes	Yes
Diffuse		Yes	Yes
Stretch	Yes		
Flow		Yes	Yes

Exercise 1.2: Properties of solids, liquids and gases

1

	Solids	Liquids	Gases
Do they flow easily?	No	Yes	Yes
Can they be compressed?	No	No	Yes
Can they change their shape?	Not easily	Yes	Yes
Are the particles close together or far apart?	Close together	Close together	Far apart
Do the particles hold on to each other tightly?	Yes	No	No

2 (a) Solid (c) Liquid (e) Solid

(b) Gas (d) Gas (f) Liquid

3 (a) The particles in the air bounce around inside the tyre; this causes pressure inside the tyre.

(b) The air pressure increases when the mechanics pump up the tyre because more particles are introduced into the same space. There are more particles bouncing around, so more pressure is created.

(c) The air pressure inside the tyre rises as it gets hotter because the particles have more (kinetic) energy and move faster. They hit each other and the walls of the tyre harden, causing an increase in pressure.

(d) Tyres that contain air absorb some of the bumps on the racing circuit because the particles can be pushed closer together.

Investigation: States of matter (pages 3–4)

1 Solid, liquid and gas

2 Thermal energy

3 (a) Melting (b) Evaporation

4 (a)

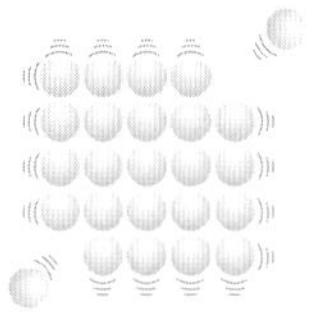

Solid to liquid transition (melting).

(b)

Liquid to gas transition (evaporation).

Investigation: Heating and expansion (page 5)

1 After heating the metal ball no longer fits through the loop. Heating has caused the metal ball to expand.

2 When the tube is warmed the liquid level rises up the capillary tube. Heating has caused the liquid to expand.

3 When the flask of air is warmed by the hands bubbles emerge from the flask. Heating has caused the air in the flask to expand.

4 Gases, liquids, solids

5 The particles move further away from one another.

6 Kinetic energy

2 Elements and compounds

Exercise 2.1: The Periodic Table

1 **Elements** are substances that cannot be broken down into simpler substances. Some, such as carbon, are made of particles called **atoms** and others, such as oxygen, are made of particles called **molecules**. There are about a **hundred** of these substances. The heaviest ones can only be made during **nuclear** reactions.

2 Carbon, magnesium, sulfur, lead

Exercise 2.2: Metals and non-metals

1 (a) **Non-metals** form much of our world.

(b) Air is mostly a mixture of nitrogen and **oxygen**. Other elements, such as **neon/argon/helium** and **neon/argon/helium**, are found in much smaller proportions.

(c) The most common element in the sea is **oxygen** (in water!), followed by **hydrogen**.

(d) Most metals are solids. The exception is **mercury**, which is liquid at room temperature. Metals are usually much tougher than non-metals, although the non-metal **diamond (allow carbon)** is the hardest natural material on the Earth.

(e) The most common difference between metals and non-metals is that metals are good **conductors** of thermal energy and electricity, whereas non-metals tend to be **insulators**.

2 (a)

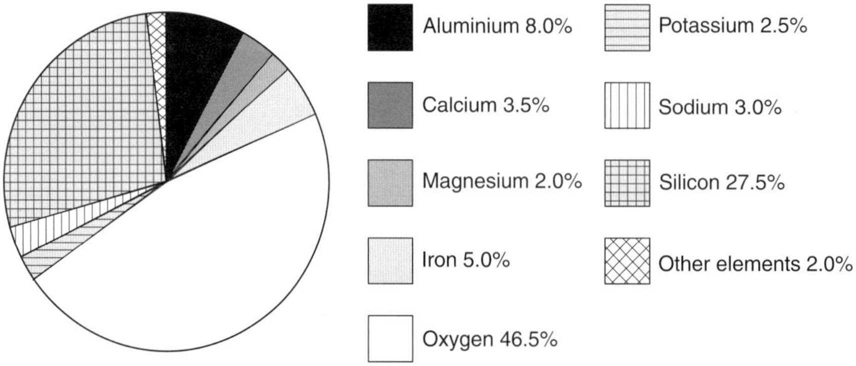

A pie chart to show the percentage by weight of different elements in the Earth's crust

(b) Aluminium

(c) In sand (silicon dioxide)

(d) Only iron is magnetic so you would use a magnet.

(e) Carbon/hydrogen

Exercise 2.3: Compounds

1

Name of substance	Chemical symbol or formula	Solid, liquid or gas	Colour	Does it conduct electricity?	Any special property
Iron	Fe	Solid	Grey–black	Yes	Magnetic
Sulfur	S	Solid	Yellow–green	No	No
Iron sulfide	FeS	Solid	Black	No	No
Oxygen	O	Gas	Colourless	No	Supports combustion
Hydrogen	H_2	Gas	Colourless	No	Burns with a pop
Water	H_2O	Liquid	Colourless	Weakly	Excellent solvent

2 (a) A compound contains two or more elements that are chemically combined. Its chemical properties are entirely different from those of the elements from which it is made.

(b) Al_2O_3, NaCl, HCl, CO and H_2O are all compounds

3 (a) Sodium, nitrogen and oxygen

(b) Magnesium, carbon and oxygen

(c) Calcium and carbon

(d) Nitrogen, hydrogen and oxygen

(e) Aluminium and oxygen

(f) Hydrogen, sulfur and oxygen

4 (a) (ii) and (vi)

(b) (i), (iii) and (vii)

(c) (iv), (v) and (viii)

(d) (ii)

(e) (i), (iii), (iv), (v), (vi) and (vii)

Extension question

5 (a) C, D (b) B (c) F

Investigation: How do compounds differ from elements? (pages 24–25)

1 The mixture retains the properties of the two elements in it, but the compound has different properties to either of the elements that it contains.

2 Iron is magnetic. Iron sulfide is not magnetic.

3 Sulfur: Sulfuric acid, sulfur dioxide, copper sulfate, for example.
Iron: Iron oxide, for example.

3 Water

Exercise 3.1: Water and the water cycle

1 Pure water boils at **100 °C** and freezes at **0 °C**. A simple chemical test uses cobalt chloride paper to test for the presence of water. The cobalt chloride paper changes from **blue** to **pink** if water is present. Seawater is a **mixture** of many different substances. The presence of impurities in seawater **lowers** the freezing point and **raises** the boiling point of water.

2

Experiment	Factor to change (independent variable)	Factor to measure (dependent variable)	Factors to keep constant (control variables)
(a)	Speed of blower	Time taken for cloth to dry	Distance/size of cloth/initial 'wetness' of cloth/type of fabric
(b)	Type of cloth	Time taken for cloth to dry	Distance/size of cloth/initial 'wetness' of cloth/speed of blower/type of fabric

Investigation: The effect of temperature and air flow on evaporation (page 30)

1 Wind moves damp air away from the washing, after water has evaporated. The thermal energy from the Sun will raise the temperature, which increases the rate of evaporation.

2 The warmer air can evaporate more water and can also hold more water vapour. The warm air flows (forms a wind) between warm areas and cooler ones – this is more violent if the temperature gradient is bigger.

4 Pure substance or mixture?

Exercise 4.1: Pure substance or mixture?

1 To check that a mixture of sand and salt actually contained two different types of particle, two approaches could be taken:

- Add water to the mixture, stir and then filter out the sand.
- Use a magnifying glass to look at a sample of the mixture. Sand is easy to distinguish from the salt crystals.

2 To check the delicatessen's claim, you would find the boiling point of the liquid. Pure water boils at 100 °C.

An alternative answer could be to evaporate the water and see if there is a deposit left behind.

3 Because more than one substance is listed and a pure substance cannot be a mixture of substances!

4 You would use a magnet. Only the iron filings would be magnetic. If the metallic powder contained magnesium particles, they would be left behind. Magnesium is not magnetic.

Exercise 4.2: Solutions and solubility

1

Word	Definition
Dissolve	What happens when one substance seems to disappear when it is mixed with a liquid
Concentrated	A solution with many solute particles in a small volume of solvent
Dilute	A solution with very few solute particles
Solute	The name for a substance that dissolves in a liquid
Soluble	This means 'can dissolve'
Solvent	The name for the liquid part of a solution
Solution	A mixture of a solvent and a solute
Insoluble	This means 'cannot dissolve'
Saturated	A solution that cannot accept any more solute
Solubility	The amount of a substance that will dissolve in a liquid

2 (a) Solutes: sugar, phosphoric acid and salt; solvent: water

(b) To make sure that the solutes dissolve quickly in the solvent they would raise the temperature and stir when the solutes are added. They could also use solutes in the form of fine powder.

3 (a) Independent (input) variable: temperature of water

(b) Dependent (outcome) variable: the amount of sugar that can be dissolved

(c) Fair test criteria: the same volume of water has been used and there has been the same amount of stirring

(d) In this instance the independent variable is lump size. The dependent variable would be the time taken for the lump to dissolve. The temperature, the volume of water and the amount of stirring are the variables that must remain fixed to ensure that this is a fair test.

Extension question

4 (a)

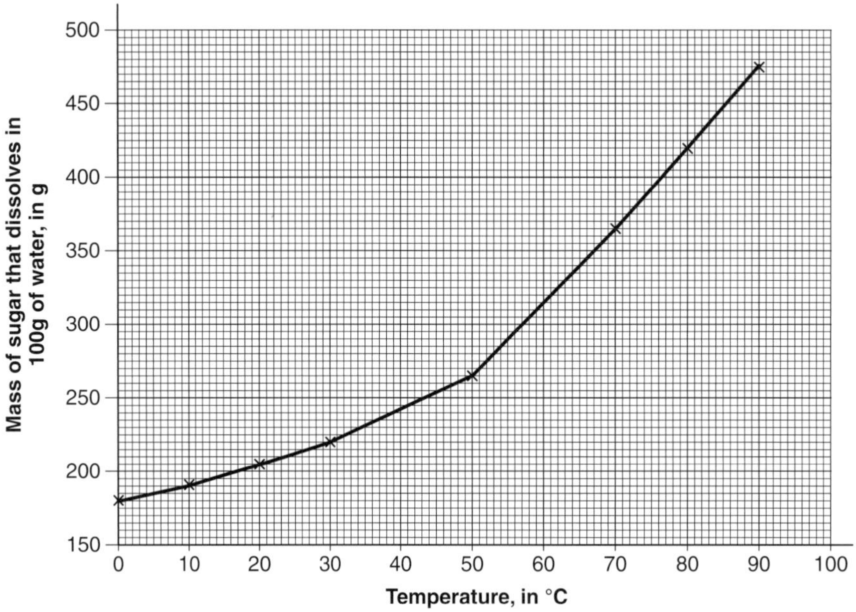

A graph to show the results of an investigation into the solubility of sugar in water

(b) 240–245 g

(c) 662.5 g

(d) Factors that they would need to keep constant if this was to be a fair test:

- the size of sugar particles
- the amount of stirring
- the volume of water.

(e) Reliability of the results could be improved by:

- repeating the experiment and taking mean (average) of the results
- using smaller increments of temperature.

Investigation: Boiling points (page 35)

1 100 °C

2 The boiling point rises if solute is added.

Investigation: Limits to solubility (page 38)

1 A saturated solution is one that contains the maximum amount of solute that can be dissolved in that volume of solvent.

2 Solubility is the maximum quantity of a substance (the solute) that may be dissolved in another (the solvent).

3 Two from: temperature, type of solvent, pressure (for gases).

5 Separating mixtures of materials

Exercise 5.1: Separating mixtures

1 (a) Most natural substances are **mixtures**, they are not pure. The particles of each substance in a mixture are not **bonded** to each other and so these substances can often be separated because they have different **physical** properties.

(b) There are several different ways of separating substances, including **chromatography**, which can separate different soluble substances in a mixture, and **evaporation**, which can provide pure crystals of a solute from a solution.

(c) The process of distillation depends on the fact that different substances have different **boiling points**. The process can be used to collect **pure water** from seawater and **alcohol** from beer or wine.

2 (a) The different substances in the sweepings can be separated from each other by first mixing with water and stirring. Only the salt would dissolve. The mixture could then be filtered. The salt could then be collected by evaporation of the salty water. The other substances would be collected in the filter paper. They could then be separated as follows:

- iron filings are magnetic so could be collected by a magnet
- sand and aluminium shavings have different appearances and might be separated by sieving (dependent on size of shavings).

More astute pupils are likely to say that sand and aluminium cannot be satisfactorily separated by physical methods.

(b) He could use anhydrous copper sulfate; if water is present it turns blue. Or he could use cobalt chloride paper; if water is present it turns from blue to pink.

3 (a) This method of separation is called chromatography.

(b) Green, brown, yellow and violet

(c) Four: E104, E110, E133, E122

(d) Orange, brown and yellow

Investigation: Separation of rock salt (page 45)

1 Sodium chloride

2 The pestle and mortar are used to grind up the rock salt lumps into smaller particles. Smaller particles dissolve more easily than larger ones.

3 The rock salt begins to dissolve in the water. The dissolving process is speeded up by stirring the solution.

4 Insoluble material, such as grains of sand.

5 A solution of sodium chloride in water.

Investigation: Food colourings chromatography (page 52)

1 (Answer depends on mixtures used.)

The choice is made because 'spots' of the substances in the mixtures rise up the paper to different extents. The two identical dyes will have spots that rise to identical positions.

2 Propanone. This solvent is used for solutes that are not soluble in water (for example, removal of nail varnish).

6 Material changes

Exercise 6.1: Changing materials

1

Diagram	Physical or chemical change?	Reason
1 A melting ice cream	Physical	No new substance formed
2 Burning a match	Chemical	Heat released/new substance formed/cannot be reversed
3 Making bread from dough	Chemical	Heat used/new substance formed/cannot be reversed
4 Stirring sugar into a cup of tea	Physical	No new substance formed/can be reversed
5 Condensation on a mirror	Physical	No new substance formed/can be reversed
6 Mercury rising in a thermometer	Physical	No new substance formed/can be reversed
7 Burning a piece of paper	Chemical	Heat released/new substance formed/cannot be reversed
8 Making glass from sand	Chemical	Heat released/new substance formed/cannot be reversed
9 Making alcohol	Chemical	Heat released/new substance formed/cannot be reversed
10 Melting gold	Physical	No new substance formed/can be reversed

Exercise 6.2: Conservation of mass

1 200 g

2 (a) (i) (81 + 75) −3 = 153 g

(ii) 105 + 111 = 216 g

(b) (i) Gas is released.

(ii) Carbon dioxide is a gas, so it escapes from the beaker – the loss in mass corresponds to the mass of the carbon dioxide.

Investigation: What happens to a solute when a solution is formed? (page 57)

1 There is no change to the mass – this is an example of conservation of mass during a physical change.

2 The particles of the solid and liquid move in between one another.

3 The student's drawing should show particles of solute (one colour) distributed among particles of solvent (different colour).

4 The mass of a solution is equal to **the mass of the solute plus the mass of the solvent that formed the solution**.

7 Chemical reactions

Exercise 7.1: Chemical changes

1 You might see fizzing, colour change and a new product being formed when a chemical change takes place. Light might be given out.

2 You may hear fizzing or popping when a chemical change takes place.

3 (a) sodium + chlorine → sodium chloride

(b) Thermal energy is given off, there is a change in colour and a different solid is formed.

4 (a) (i) magnesium + oxygen → magnesium oxide

(ii) Magnesium oxide has more mass than magnesium. The extra mass is oxygen from the air.

(b) Oxygen

(c) Zinc oxide

(d) A – chemical; B – chemical; C – physical

Exercise 7.2: Important chemical changes

1

Time/minutes	Volume of dough/cm^3
0	24 (allow 25)
5	32 (allow 33, 34)
10	42 (allow 43)
15	64 (allow 65)
20	72 (allow 73)
25	72 (allow 73)
30	72 (allow 73)

(a)

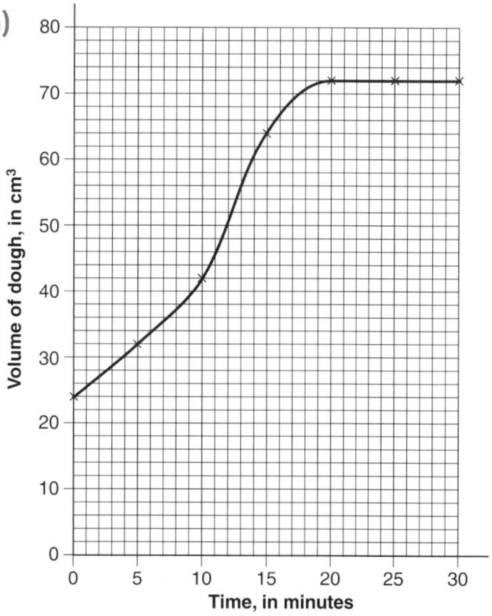

A graph to show how the amount of dough changes over the half-hour period

(b) 11–12 minutes

(c) (i) The independent (input) variable would be the amount of sugar.

(ii) The dependent (outcome) variable in this experiment would be the volume of dough.

(iii) For this experiment to be a fair test the baker would need to control the temperature (by using a water bath and a thermometer or by placing all the measuring cylinders in the same (warm) place) and the initial volume of dough (by using a measuring cylinder).

2 Check pupils' answers. Examples could be:

- useful: fermentation, respiration, photosynthesis, digestion
- not useful: rusting, decay, weathering of buildings

3 Check that pupils have selected ten substances that could be found in their homes. Check they have identified any chemical reactions correctly and that they have found out which reactants were needed to make each one of the substances.

Exercise 7.3: Conservation of mass

1 From the results of this experiment we can make several observations. Firstly, note that the mass of the crucible remains the same, 50 g. This allows us to calculate what happens to the mass of the contents before and after heating. The mass of the magnesium ribbon before heating is 12 g. After heating, the contents of the crucible (now magnesium oxide) have increased in mass to 20 g. The additional mass of 8 g is the result of the combination of oxygen from the air with the magnesium ribbon to form magnesium oxide.

Exercise 7.4: Burning

1 Putting a blanket over burning wood prevents the supply of the oxygen needed for combustion, and so helps to extinguish the fire.

2 (a)

Type of fuel	Units of thermal energy released	Amount of fuel burned/grams	Units of thermal energy from 100 grams of fuel
Coal	40	60	66.7
Gas	54	80	67.5
Paraffin	36	50	72
Petrol	60	50	120
Diesel oil	54	75	72

(b) It is important to complete the final column to allow a like-for-like comparison (i.e. you are comparing the performance of a fixed mass of fuel).

(c) The same amount of oxygen must be available in each case otherwise the experiment would not be a fair test. Therefore the results would be unreliable.

(d)

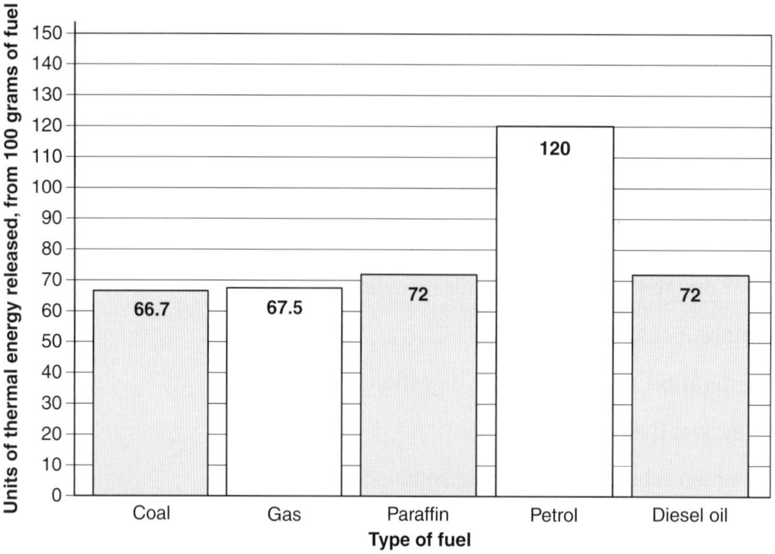

A bar chart to show how much energy is given out when fuel burns

(e) On the evidence of this experiment, petrol would seem to be the most efficient heating fuel.

(f) Despite the evidence suggesting that petrol is the most efficient heating fuel, petrol is very expensive and difficult to store and deliver, so it is not the most useful heating fuel. Diesel oil and paraffin are the next most efficient fuels, but these also have problems of storage and delivery, as does coal. Gas would seem to be the most practical/useful heating fuel due to its method of piped supply, even though its efficiency is slightly less than that of diesel oil and paraffin.

Exercise 7.5: Air pollution

1 Wood

2 We say that carbon dioxide is a 'greenhouse' gas because it acts like the glass in a greenhouse. It keeps the thermal energy close to the Earth's surface and so causes warming.

Extension questions

3 (a) (b)

Concentration of sodium disulfite/%	Number of seeds germinated out of 25 experiments	Mean number of seeds germinated out of 50 planted	Mean percentage germination
0.00	19, 19, 17, 20, 18	18.6	37.2
0.05	18, 19, 18, 19, 19	18.6	37.2
0.10	12, 13, 14, 11, 12	12.4	24.8
0.50	0, 1, 0, 0, 1	0.4	0.8
2.50	0, 0, 0, 0, 0	0	0

(c) 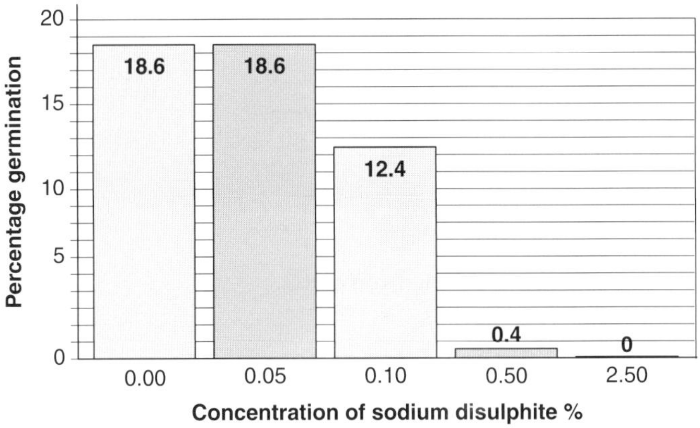

A bar chart to show the effects of sodium disulfite on the germination of oat seeds

(d) The experiment was repeated five times to enable the students to check the reliability of the results; whether they are all the same within the bounds of experimental accuracy. Having repeated the results, the mean can be calculated in order to provide a single figure to work with. This will be as reliable as possible. It also enables students to increase accuracy by repeating any experiments that produced anomalous results.

(e) A control is used to show the germination rate when water is available but no sulfur dioxide. This result can then be compared with the others to determine the effect of the sulfur dioxide.

(f) 0.10%

(g) They could reduce the increments of sodium disulfite between 0.05 and 0.10% to get a more accurate value.

(h) Independent variable – concentration of sodium disulfite; dependent variable – percentage germination. Control variables might be temperature, availability of water and the availability of oxygen.

(i) (i) Burning fossil fuels, e.g. in thermal power stations, releases large amounts of sulfur dioxide into the natural environment. Industry produces most of the SO_2. Cars etc. produce carbon and nitrogen oxides because most petrol is low sulfur these days.

(ii) Other effects could be: irritation of lungs, reduced photosynthesis, acidification of lakes preventing growth of crustaceans.

4 (a) 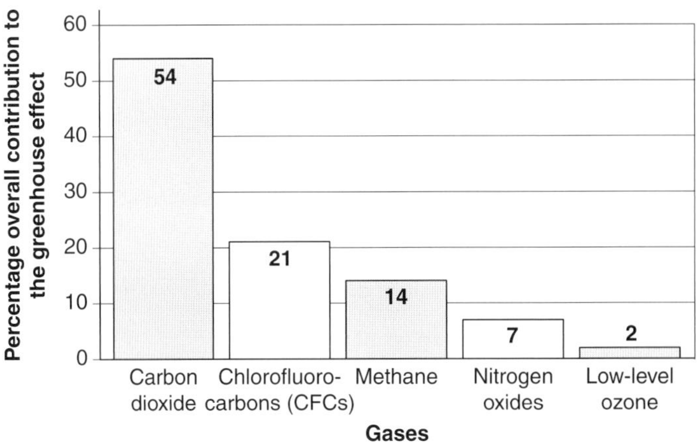

A bar chart to show the sources and effects of greenhouse gases

(b) Effect of water 2%

(c) Carbon dioxide and methane

(d) A greenhouse gas acts like the glass in a greenhouse. It keeps thermal energy close to the Earth's surface and so causes warming.

(e) Harmful effects of greenhouse gases: melting of ice caps, flooding, spread of pests, altered weather patterns, including storms.

(f) (i) Less carbon dioxide from burning and less methane from cattle. Forests are big users of CO_2 during photosynthesis.

(ii) Less burning of fossil fuels to provide thermal energy for houses.

(iii) Less burning of fossil fuels to provide electricity.

(g) (i) 0.005%

(ii) 0.020%

(iii) A vast increase in the number of cars in the world has led to a much greater use of fossil fuels.

Exercise 7.6: Reversed reactions

1 No. The particle theory explains why, during any chemical reaction, there is conservation of mass. This means the total mass of the reactants used is the same as the total mass of the products formed.

2 Lime water would turn milky.

3 Calcium oxide is dissolved in water to form calcium hydroxide (slaked lime), which can then be used to neutralise acid soils. This increases the yield of many crops, as most mineral ions are absorbed more easily from neutral or slightly alkaline soils than from acid soils.

Investigation: Burning (page 71)

1 Check student's calculation. Answer will vary depending on the fuel used.

2 Carbon in the wax of the candle combines with oxygen to form carbon dioxide. This is a gas so is lost from the candle. Therefore, the mass of the candle is reduced.

Investigation: Mass change on heating (page 83)

1 0.44 g

2 The mass of the contents increases because magnesium has combined with oxygen in the air to form a new compound, magnesium oxide.

3 Conservation of mass: In a chemical reaction, the total mass of the reactants used is the same as the total mass of the products formed.

8 The reactions of metals

Exercise 8.1: The reactivity series

1 (a) Zinc is more reactive.

(b) zinc + silver nitrate → **zinc nitrate + silver**

2 (a) Iron heated with copper oxide: very slow reaction (iron close to copper in reactivity series) but iron oxide and copper will be formed.

(b) Magnesium placed in dilute hydrochloric acid: the reaction occurs quickly. Hydrogen is given off and can be tested with a lighted splint; a (squeaky) pop should be heard.

(c) Copper placed in dilute sulfuric acid: no reaction takes place because copper is lower than hydrogen in the reactivity series.

(d) Magnesium placed in copper sulfate solution: the magnesium displaces the copper from the copper sulfate to form magnesium sulfate and copper.

(e) Silver warmed with water: at this level it is fair to say that no reaction takes place as silver is lower than hydrogen in the reactivity series.

Extension question

3 (a) You could test to see what metal M is by carrying out a series of experiments, where the metal M is added to different solutions of metal salts. You would look for any displacement reactions that may take place between the unknown metal and the solutions provided. Metals displace one another in regular order, so you can use the evidence of the reactions to place the unknown metal in the reactivity series and, thus, work out what it is.

(b) The gas given off is hydrogen and can be tested with a lighted splint – a (squeaky) pop should be heard.

(c) The gas is carbon dioxide. Its identity could be tested by bubbling it through limewater. If it is carbon dioxide, the limewater will turn milky.

Exercise 8.2: Corrosion

1 (a) Corrosion involves a reaction between a **metal** and some substance in the **environment/atmosphere**. In most cases an **oxide** is formed on the surface of the metal.

(b) Rusting is the corrosion of **iron** and **steel**. This is a dangerous process because the **rust/hydrated iron oxide** is weak and brittle. Rusting can be prevented by coating the metal with, for example, **plastic/paint/zinc**. Another method of prevention involves 'sacrificing' a second metal, such as **zinc/magnesium**.

2 **(a)** iron + water + oxygen → hydrated iron oxide

(b) Cars rust more quickly in the UK than in California because there is more water in the atmosphere in the UK.

(c) By painting or galvanising

3 **(a)** Galvanising is such an effective method of protection because it offers two levels of protection: barrier and sacrificial.

(b) Examples could be: buckets, baths, fence posts, cars and bridge supports.

(c) Coating cannot be used for preventing corrosion of railway lines because a coating would be worn away by the abrasion of locomotive wheels.

4 **(a)** Warmth/temperature had the bigger effect on rusting. This can be deduced by comparing tubes 3 and 4 with tubes 1 and 2.

(b) The fifth tube: 0. Assume that the boiled water will have no oxygen, and no oxygen can enter from the air because of the oil layer. No rusting is possible without air.

(c)

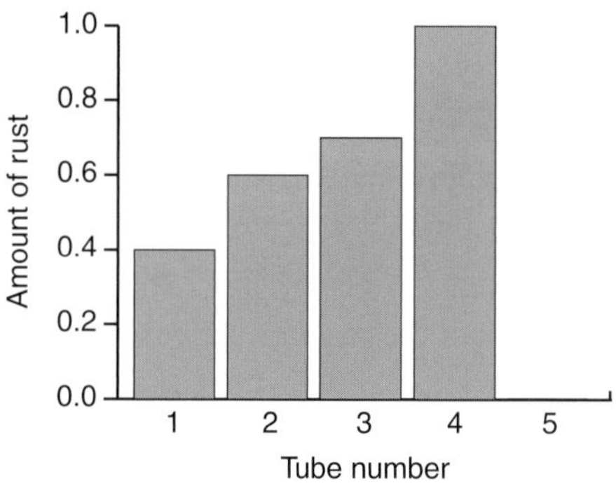

A chart to show the results of a rusting experiment

Extension questions

5 Design of a fair test to investigate whether galvanising offers double protection for steel.

- Independent variable will be level of protection.
- Dependent variable will be extent of rusting.
- Method: Compare the extent of rusting seen on a galvanised nail with a scratched galvanised nail and a bare steel nail.
- Controlled variables might include temperature, access to oxygen, water and time before the results are collected.

6 Check pupils' investigations into stainless steel. Check for mentions of how it is different from iron: stainless steel contains iron, nickel and chromium. The alloy of these metals does not rust because a surface layer of chromium oxide forms and prevents damp air getting to the iron. The layer is very strong, despite being very thin – less than 0.000 000 01 m thick.

7 (a)

Mass of magnesium/g	Starting temperature/°C	Final temperature/°C	Rise in temperature/°C
0.00	22	22	0
0.25	23	30	7
0.50	23	38	15
0.75	22	46	24
1.00	22	55	33
1.25	22	61	39
1.50	23	68	45
1.75	24	69	45
2.00	23	68	45
2.25	22	67	45
2.50	23	68	45

(b) copper sulfate + magnesium → copper + magnesium sulfate

(c) The independent (input) variable for this reaction is the mass of magnesium.

(d) The dependent (outcome) variable in this experiment is the rise in temperature.

(e) Exothermic

(f)

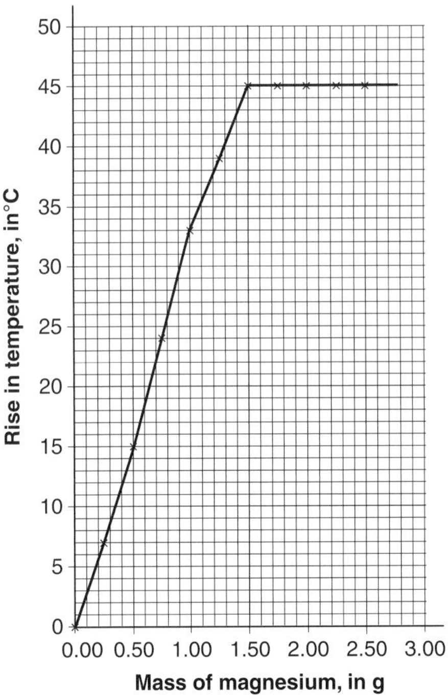

A graph to show the results of an investigation into how much change in thermal energy went on when magnesium reacted with copper sulfate solution

(g) The curve increases (i.e. the temperature rises) as the amount of magnesium is increased. At 1.50 grams the curve levels off because at this stage all the copper sulfate has reacted and the reaction ceases. No more rises in thermal energy are recorded after this point.

(h) Fair test criteria: fixed amount of copper sulfate; a fixed starting temperature and a fixed volume of vessel

(i) Total mass at the end of the experiment is 26.5 g. This is because during this chemical change there is conservation of mass. This means the total mass of the reactants used is the same as the total mass of the products formed.

Exercise 8.3: Extraction of metals

1 Some metals, such as **gold** and silver, are found as the uncombined metal in nature. Most metals are found combined as **compounds/oxides** in **ores** such as hematite and bauxite. There are three stages in the extraction of a metal: **mining**, **decomposition** (which always involves some chemical reactions) and **purification** (which makes the metal suitable for use).

2

Metal	Main ore	One important use of the metal	Method of extraction
Gold	Found native	Jewellery/electrical contacts	Mining
Iron	Hematite	Building materials/ any suitable example	Heating with carbon
Copper	Malachite	Wire for the conduction of electricity	Direct heating/ heating with carbon
Aluminium	Bauxite	Cooking utensils/ cars/aircraft	Electrolysis
Mercury	Cinnabar	Thermometers	Heating in air

3 (a) mercury oxide → mercury + oxygen

(b) tin oxide + carbon → tin + carbon dioxide

Investigation: The rusting of iron (page 97)

1 Conclusions: Both air and water are needed for iron to rust. Rusting occurs more quickly in the presence of acid or salt. Rusting occurs more quickly at higher temperatures.

9 Acids, bases and indicators

Exercise 9.1: Acids and bases

1 Pickling foods in vinegar provides an acid environment that prevents the multiplication of most bacteria.

2 Check pupils' answers. Examples could be nitric acid in production of fertilisers and explosives; sulfuric acid in making fertilisers, paints, plastics and in car batteries; hydrochloric acid in the processing of metals and purification of ores.

3 An alkali is a soluble base/a soluble substance that can neutralise an acid. Check pupils' answers. Many alkalis are used in cleaning products (e.g. oven cleaner, toothpaste, soap).

4 Acids and alkalis are **corrosive**, which means that they can cause damage to the skin. If one of these substances is spilt or splashed onto the skin, plenty of cold water must be run over the splashed area. In the laboratory, you should always add **acid** to **water** and never the other way round. When working with acids or alkalis, you should always wear **overalls/lab coat** and **goggles/eye protection**.

Extension questions

5 Hydrochloric acid helps in a number of ways: it provides ideal conditions for the enzyme involved in the digestion of protein in the stomach; it kills harmful bacteria in food. However, hydrochloric acid can be harmful because it can damage the lining of the stomach (causing ulcers) or the base of the gullet (causing heartburn).

6 Check pupils' answers. An antioxidant 'mops up' oxidising agents. These oxidising agents can damage proteins and DNA in cells. Antioxidants are abundant in brightly coloured foods e.g. peppers, broccoli and tomatoes.

Exercise 9.2: Neutralisation

1 To make a nettle-sting less painful you would rub on a weak alkali (such as a solution of baking powder). Rubbing with a dock leaf is another commonly known treatment, but it has been found that dock leaves are also acidic and, in fact, their soothing affect comes from the moist, cooling sap.

2 This shows that the aspirin solution is an acid.

3 (a) D (b) B, C, E (c) A (d) B

Extension questions

4 The antacids would be the independent (input) variable. The amount of antacid required to neutralise an acid would be the dependent (outcome) variable. Control variables (to make it a fair test) would be the volume of acid used and the type of acid used. The temperature during the experiment would also remain fixed. Remember that the most powerful antacid will require the smallest amount to neutralise the acid.

The apparatus needed for this experiment would include:

- pestle and mortar for grinding up the tablets
- weighing machine to determine fixed mass of remedy
- four beakers (or conical flasks) each containing 10 cm^3 of acid; beakers labelled, for example A, B, C, D
- water to dissolve remedies
- four 20 cm^3 syringes containing the four antacids to be tested; syringes labelled, for example A, B, C, D
- test-tube rack if test tubes are used instead of beakers
- full-range or universal indicator to determine point of neutralisation
- white tile or plain white paper to place under testing flask/beaker
- eye protectors.

The method would be:

- set up equipment needed
- place 10 cm^3 (fixed value for each) of acid in each beaker/flask
- add 2–3 drops of indicator and make a note of the colour
- grind up each antacid tablet and dissolve in a fixed volume of water; separate beakers labelled A, B, C, D
- use syringe for first antacid solution and add (2 cm^3 at a time) to acid in beaker, swirling after each addition until neutral green is obtained
- record volume of antacid used to neutralise the acid
- repeat the process for the remaining three antacid remedies
- compare the recorded volumes of each antacid used to neutralise the acid and reach a conclusion to the experiment.

5 (a) A graph to show the results of a neutralisation reaction

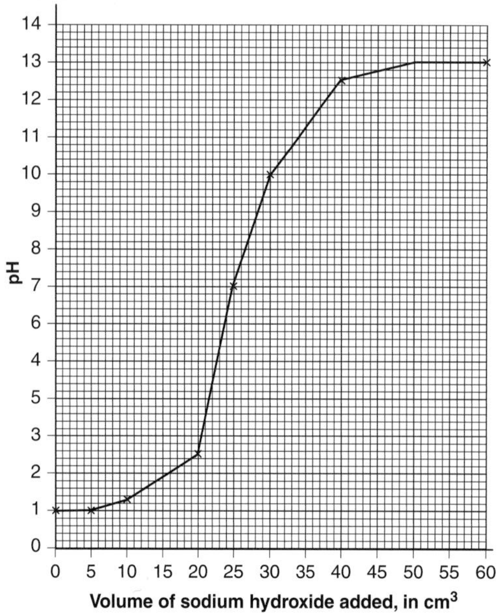

(b) Check pupils' answers. Make sure they demonstrate an understanding of neutralisation and the point at which neutralisation occurs.

(c) They could improve their results by:

- repeating the experiment several times and using the mean result
- using smaller increments of volume of sodium hydroxide added, particularly close to the neutralisation point.

Exercise 9.3: More reactions of acids

1 (a) zinc + **hydrochloric acid** → zinc chloride + hydrogen

(b) nitric acid + magnesium → **magnesium nitrate** + **water**

(c) **sulfuric acid** + potassium hydroxide → potassium sulfate + **water**

(d) copper carbonate + hydrochloric acid → **copper chloride** + **carbon dioxide** + **water**

(e) lead + **sulfuric acid** → lead sulfate + **water**

2 (a) Copper chloride: hydrochloric acid and copper oxide

(b) Lead nitrate: nitric acid and lead oxide

(c) Iron chloride: hydrochloric acid and iron oxide

(d) Zinc sulfate: sulfuric acid and zinc oxide

3 The acid could react with the metal, but would not react with glass.

4 (a) You can tell that this is a chemical change because a gas is released.

(b) Carbon dioxide is the gas given off. Carbon dioxide turns limewater milky.

(c) You can tell when the reaction is finished because no more gas is given off.

(d) Calcium chloride

(e) Evaporate the solution very slowly.

5 Acids react with most metals to produce a **salt** and a gas called **hydrogen**. This gas makes a **(squeaky) pop** when tested with a lighted splint. Acids react with **carbonates** to make a salt, water and carbon dioxide gas. Limestone contains the compound **calcium carbonate**, which can be dissolved by acid in rainwater.

Extension questions

6 The acid in the fruit will react with iron but not with copper.

7 (a) A graph to show the results of an experiment to investigate the reaction between marble chips and dilute hydrochloric acid

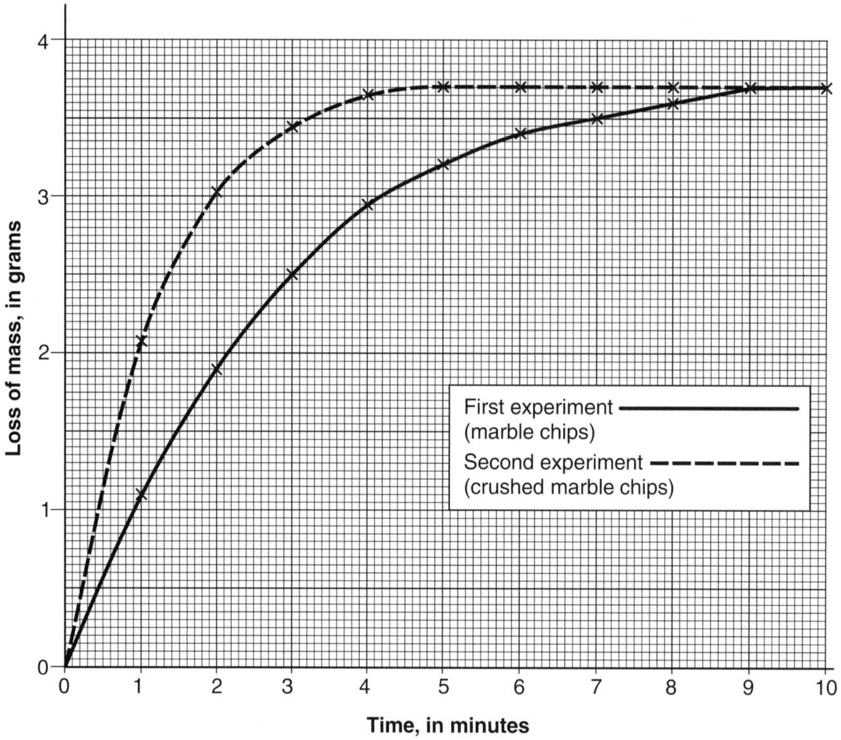

(b) The second experiment goes the fastest at the start of the reaction.

(c) The reaction has finished. There is a finite mass of hydrochloric acid and at this stage there are no more molecules of hydrochloric acid to react with the marble chips. No more calcium chloride can be produced; no more carbon dioxide will be released. The mass becomes constant after 9 minutes.

(d) An increased surface area increases the rate of a chemical reaction.

Investigation: Indicators (page 115)

1 Acids: hydrochloric acid, lemon juice and shampoo. Alkalis: sodium hydroxide, limewater and tapwater (tapwater will vary by locality).

2 Universal indicator shows a better graduation of pH. Litmus only shows whether a sample is acid or alkali.